This report contains the collective views of an international group of experts and does not necessarily represent the decisions or the stated policy of the World Health Organization or of the Food and Agriculture Organization of the United Nations.

Strategies for assessing the safety of foods produced by biotechnology

Report of a Joint FAO/WHO Consultation

World Health Organization
Geneva 1991

WHO Library Cataloguing in Publication Data

Joint FAO/WHO Consultation on the Assessment of Biotechnology in Food Production and Processing as Related to Food Safety (1990: Geneva, Switzerland)
  Strategies for assessing the safety of foods produced by biotechnology: report of a joint FAO/WHO consultation.
  1. Biotechnology – standards – congresses 2. Food technology – standards – congresses I. Title

ISBN 92 4 156145 9   (LC Classification: TP 248.14)

© WORLD HEALTH ORGANIZATION 1991

Publications of the World Health Organization enjoy copyright protection in accordance with the provisions of Protocol 2 of the Universal Copyright Convention. For rights of reproduction or translation of WHO publications, in part or *in toto*, application should be made to the Office of Publications. World Health Organization, Geneva, Switzerland. The World Health Organization welcomes such applications.

The designations employed and the presentation of the material in this publication do not imply the expression of any opinion whatsoever on the part of the Secretariat of the World Health Organization concerning the legal status of any country, territory, city or area or of its authorities, or concerning the delimitation of its frontiers or boundaries.

The mention of specific companies or of certain manufacturers' products does not imply that they are endorsed or recommended by the World Health Organization in preference to others of a similar nature that are not mentioned. Errors and omissions excepted, the names of proprietary products are distinguished by initial capital letters.

TYPESET IN INDIA
PRINTED IN ENGLAND

91/8871-Macmillan/Clays-6000

Contents

1. **Introduction**   1
   1.1 Scope of the Consultation   2
   1.2 History of the use of biotechnology in food production   3

2. **Applications of biotechnology in food production and processing**   6
   2.1 Bacteria and fungi   6
       2.1.1 Fermented foods   6
       2.1.2 Food additives and processing aids   7
       2.1.3 Applications using enzymes   9
       2.1.4 Products used in agriculture   9
   2.2 Plants   11
   2.3 Animals   14
   2.4 Food analysis   16

3. **Safety assessment of foods derived from micro-organisms generated by biotechnology**   18
   3.1 Introduction   18
   3.2 Issues to be considered in safety assessment   19
       3.2.1 General considerations   20
       3.2.2 Specific safety assessment considerations   21
   3.3 Safety assessment paradigm   28
   3.4 Summary   29

4. **Safety assessment of foods derived from plants generated by biotechnology**   30
   4.1 Introduction   30
   4.2 Issues to be considered in safety assessment   31
       4.2.1 General considerations   31
       4.2.2 Specific safety assessment considerations   35
   4.3 Safety assessment paradigm   37
       4.3.1 Animal studies   38
       4.3.2 Human data   38
   4.4 Summary   39

5. **Safety assessment of foods derived from animals generated by biotechnology** — 40
   5.1 Introduction — 40
   5.2 Issues to be considered in safety assessment — 40
      5.2.1 Gene products — 41
      5.2.2 Genetic construct — 41
      5.2.3 Unintended genetic effects — 42
   5.3 Safety assessment paradigm — 43
   5.4 Summary — 43

6. **Recommended safety assessment strategies for foods and food additives produced by biotechnology** — 45
   6.1 Introduction — 45
   6.2 General considerations — 45
      6.2.1 Biological characteristics — 46
      6.2.2 Molecular characteristics — 46
      6.2.3 Chemical characteristics — 46
   6.3 Specific recommendations — 47
      6.3.1 Safety assessment of genetically modified microorganisms and foods produced by them — 47
      6.3.2 Safety assessment of genetically modified plants and foods derived from them — 48
      6.3.3 Safety assessment of genetically modified animals and foods derived from them — 49

7. **Conclusions and recommendations** — 50
   7.1 Conclusions — 50
   7.2 Recommendations — 52

References — 53

Annex 1. List of participants — 55

Glossary — 57

1. Introduction

A Joint FAO/WHO Consultation on the Assessment of Biotechnology in Food Production and Processing as Related to Food Safety was held in Geneva from 5 to 10 November 1990; the participants are listed in Annex 1. The Consultation was opened by Dr J. Rochon, Director, Division of Health Protection and Promotion, WHO, on behalf of the Directors-General of FAO and WHO. In welcoming the participants, Dr Rochon drew attention to the long history of the application of biotechnology to food production and processing. This went back more than 8000 years, so that the food industry was one of the oldest users of biotechnological products and processes.

Since the nineteenth century, the science of biotechnology had developed more rapidly than ever before and particularly so over the past decade. While there were many applications of biotechnology in areas such as drugs, the new technologies were also potentially capable of revolutionizing the world's food supply. Enormous improvements were possible in both the quantity and quality of food available. Contemporary techniques of genetic modification made it possible both to speed up the classical processes of plant and animal breeding and to effect interspecies gene transfers not possible by classical methods.

Dr Rochon predicted that biotechnology, by changing the character of food sources, would have an enormous impact on our ability to provide food for the world's rapidly increasing population. The challenge would be to develop appropriate safety assessment procedures to ensure that these new food sources were safe for human consumption. An international consensus on the safety assessment of foods derived from biotechnology would be a strong foundation for consistent national regulatory activities and it was hoped that the Consultation would be a valuable initiative in that direction.

Biotechnology raised a number of important nonscientific issues related to ethics, consumer perceptions and food labelling, which would need to be taken into account by national regulatory agencies. Such issues were, however, outside the remit of the Consultation, which would confine its activities to the scientific issues of safety assessment.

In his reply, Dr S. A. Miller, Chairman of the Consultation, emphasized the need for a strong science base for any national and international regulatory activities if biotechnology were to move forward. A careful balance would be necessary to ensure that any possible problems were neither understated nor overstated; it was important that the public should not be unduly alarmed, and equally that any gaps in available information should be acknowledged.

1.1 Scope of the Consultation

The aim of the Consultation was to outline appropriate strategies and procedures to assist those responsible for assessing the safety of specific applications of biotechnology in food production and processing. It adopted a number of definitions established by other international bodies; these, together with definitions of other terms used in this report, are given in the Glossary (p. 57). The Consultation noted that the definition of biotechnology applied equally to classical and modern techniques, underlining the fact that, from the point of view of safety, there was no fundamental difference between traditional products and contemporary ones obtained by means of biotechnology. It agreed, therefore, that the same broad principles of safety assessment should apply to the products of both the old and the new biotechnologies. The Consultation also noted that, as biotechnology was developing rapidly, there would be a need to re-examine the issues at a later date. The Consultation's report should therefore be seen as a first step in a series of activities aimed at reaching an international consensus and providing guidance on the safety assessment of foods obtained using contemporary techniques of biotechnology.

In addition to its direct application to food production and processing, biotechnology also has applications in the production of veterinary drugs, pesticides and other products used in agriculture, and also in the development of improved methods for use in food analysis. Such aspects, which are of significance in food safety, are briefly considered in sections 2.1.4 and 2.4.

While environmental aspects of biotechnology are also important, they are not dealt with directly in this report as they are being dealt with elsewhere; e.g., both the United Nations Environment Programme and the Organisation for Economic Co-operation and Development are active in this field. The notification requirements that apply to the deliberate introduction into the environment of genetically modified food organisms, whether microbes, plants or animals, will usually have to be satisfied before any consideration is given to food safety. Many of the data necessary for the examination of these environmental aspects will also be of value in assessing food safety, in particular, information on the characteristics of the host and donor organisms, genetic insert and vector. Environmental information on genetically modified plants is particularly relevant to food safety as it also covers the potential transfer of genetic material to other food crops through pollen transfer.

1.2 History of the use of biotechnology in food production

Biotechnology has been used in food production and processing for thousands of years. Almost every ingredient used in the production of food has as its source a living organism, whether animal, plant or microorganism. The food sources available to early humans, both plant and animal, had evolved through natural selection. Genetic diversity arising from spontaneous genetic changes, including recombination, mutation and reproductive isolation, was exploited when the early farmers began to save the seeds from their best crops for later sowings and to use the best animals for breeding.

Even before the laws of segregation and independent assortment were formulated by Mendel in the 1860s and rediscovered in the early part of this century, their significance was recognized and applied empirically in selective breeding programmes. The simple procedure of maximizing the reproductive efficiency of organisms with advantageous phenotypes, while limiting the reproduction of those with undesirable traits, led to great improvements in the productivity of many animal and plant species. With the recognition of the validity of Mendel's laws, these practices were applied in a more scientific manner, to develop new strains of organisms for use in food production and processing. Although considerable advances have been made, there are limits to what can be achieved by conventional breeding and strain selection. The process is slow and limits are imposed by the genetic diversity of the parent organisms; results are often unpredictable and lengthy back-crossing may be necessary to remove undesirable traits introduced together with the desirable ones.

The value of breeding and selection techniques can be improved by increasing the diversity within the gene pool of the parent organisms. Muller et al., working with *Drosophila* in the 1970s, showed that X-rays could have a point effect on a single gene within the organism (*1*). Other forms of radiation and certain chemicals have been found to have a similar effect. Mutants induced by these treatments have been used successfully in plant breeding, with large numbers of progeny being produced from plants grown from treated seed. Screening identifies useful mutants which can be incorporated into breeding programmes.

The plant breeder's task has been eased considerably in recent years by the inclusion in breeding programmes of plant cell-culture techniques. In food plants, many of the traits that are the targets for change are controlled by more than one gene, and it will be many years before these systems are fully understood and can be genetically modified. The strategy of locating quantitative trait loci using restriction fragment length polymorphisms is based on the fact that DNA sequence polymorphisms, reflected as alterations in the distribution of restriction endonuclease cleavage sites, can often be identified either within or very close to a gene of interest. In the tomato, for example, genes affecting fruit mass can be identified by crossing a tomato with high fruit mass with another variety of tomato that has a number of polymorphic differences in restriction enzyme sites. Progeny plants can then be screened with probes derived from individual plant chromosomes, and the presence of specific chromosomes from the parent with high fruit mass compared with the appearance of the phenotype. In this way, traits can rapidly be mapped to the one or more chromosomes on which the relevant genes reside. This powerful new technique could be used in the rapid selection of superior plants in traditional plant-breeding programmes. Such mapping can also be used in conjunction with molecular cloning procedures to clone the various genes of interest.

In animal breeding also, new techniques have brought about significant advances. Administration of hormones to increase ovum production, followed by artificial insemination, can provide up to a dozen or so embryos which can then be implanted into surrogate mothers. In a more recent development, ten or more cells may be taken from each fertilized egg before cell differentiation, and their nuclei implanted into unfertilized eggs from which the nuclei have been removed. Each embryo is then implanted into a surrogate mother to produce identical offspring. Thus, in theory, several hundred offspring can result from a single mating, considerably reducing the time necessary to introduce a new strain. However, even these developments are limited by the inherent genetic endowment of each species and by the lack of methods for determining the structure and function of individual genes.

New methods based on molecular biology have aroused considerable interest, because they offer the prospect of more rapid and precisely targeted genetic changes than can be introduced through breeding and selection. They are also not bound by the genetic diversity of the parent organisms or their sexual compatibility. These new methods have been made possible by a number of major discoveries over the past 50 years which have built on the pioneering work of Darwin and Pasteur. They include the demonstration that DNA is the bearer of genetic information, the elucidation of the structure of DNA, and Cohen et al.'s demonstration in 1973 that DNA could be constructed *in vitro*, and inserted and expressed in a living organism (2). Organisms can thus be genetically modified by the introduction of novel genetic material in the form of a DNA construct made *in vitro*. A number of techniques can be used for this purpose, including sexual crossing, protoplast fusion and direct gene transfer.

To be of any practical value, the novel DNA introduced into the cells of the host organism must be genetically stable and properly expressed. Expression of the gene products, if achieved, will reflect precisely the nature of the modification made, although the effects of the gene products on the metabolism of the organism may not be easy to predict, particularly if the foreign gene comes from an unrelated species.

2. Applications of biotechnology in food production and processing

The status of biotechnology in food production and processing is reviewed in this section. For convenience, microorganisms, plants and animals are considered separately although many of the principles involved are similar for all three. Particular attention is given to the newer methods of biotechnology.

2.1 Bacteria and fungi

In this section, the history and current status of the use of microorganisms and their products in food production are discussed. Also included, for the sake of completeness, is a discussion of products used in agriculture, including veterinary drugs, biological pesticides and rhizobia.

2.1.1 Fermented foods

Throughout the world, fermented foods form a major part of the human diet; however, it is impossible in this report to discuss any of them in detail. Examples of fermented plant products include alcoholic drinks, tea, coffee, bread, sauerkraut, miso and tempeh. A wide variety of fermented fish, milk and meat products is also available. Fermentation, which may be brought about by yeasts, moulds or bacteria, not only helps to diversify the diet but can also contribute to increased palatability, acceptability, nutritional value and shelf-life of foods.

Many fermented foods have been known for hundreds, if not thousands, of years. Initially they were made in the home or at local level, and the strains of organisms used were passed down from generation to generation and were selected for desirable characteristics, such as flavour production. Many fermented foods are now produced on an industrial scale, and there is interest in applying genetic modification techniques to the microorganisms that produce them, including those involved in bread and beer production. The yeasts involved are well researched and potentially easy to modify; developments under way include:

(a) the incorporation into commercial brewing strains of genes for glucoamylase production, thus avoiding the need to add exogenous enzymes during beer production;
(b) the incorporation into commercial bread-making strains of a more efficient system for metabolizing maltose, thus reducing bread-making time;
(c) the incorporation into commercial yeast strains of genes to enable pharmaceutical proteins to be produced after the yeast has been recovered from food production.

In the dairy industry, lactic-acid-producing bacteria, genetically modified to produce strains with improved phage resistance or bacteriocin or flavour production, are being developed.

In the early 1970s, single-cell (i.e., bacterial or fungal) protein was thought to have considerable potential as a source of human and animal food. One project that has become a commercial success produces human food from the organism *Fusarium graminearum* grown on hydrolysed starch. The bacterium *Methylophilus methylotrophus* has also been used to convert methanol into animal feed protein, but the project has been terminated because of economic difficulties.

The most abundant renewable biomass on earth is cellulose, an estimated 5–15 tonnes being produced per capita per year. Much of the cellulose is bound physicochemically to lignin. Some higher fungi, including some edible species, can be used to convert lignocellulose directly into fungal protein suitable for human consumption.

Although there is a considerable body of literature on the production of fats and lipids by microorganisms, microbial lipids are not being produced commercially for use in food at present. This situation could change, however, with the growing consumer demand for "speciality" oils and the possible use of genetic modification, in the long term, to manipulate the fatty acid composition of microbial lipids.

2.1.2 Food additives and processing aids

A wide range of food additives, including amino acids, citric acid, vitamins, enzymes and polysaccharides such as xantham gum, has been produced for many years using microorganisms. For some products, the use of microorganisms has replaced chemical synthesis or extraction from animal or plant sources. Thus, for example, microbes have largely replaced lemons as a source of food-grade citric acid. Microorganisms have advantages as compared with animal or plant sources in terms of continuity of supply of the product and often also in terms of ease of product recovery. Products obtained from microorganisms will differ in their spectrum of impurities as compared with the corresponding chemically produced product.

Traditionally, the production of food additives from microorganisms has been optimized by the correct choice of fermentation parameters and the use of selected high-yielding strains. Many strains of microorganism used in the commercial production of food additives have been improved by processes involving radiation or mutagenic agents. Thus, for example, strategies for maximizing the efficiency of bacterial production of amino acids include inducing the bacteria to excrete glutamate by excluding biotin from the nutrient medium.

There is considerable potential for the exploitation of the new techniques of genetic modification in the production of food additives. They are already being used in a small number of instances both to increase yields and reduce processing costs of existing additives (e.g., L-tryptophan) and to develop new sources of existing additives or processing aids (e.g., chymosin). Future possibilities include the production, in microorganisms, of certain flavour substances currently synthesized chemically or obtained from plant sources.

Although plant cells are not microorganisms, their culture in fermenters is analogous to microorganism fermentation. Plant cell-culture techniques have improved dramatically over the past five years but there are still some technical problems. If these can be overcome, the rapid screening and strain improvement possible with plant cells suggest that such cells grown in culture have considerable potential for use in the production of food additives. In particular, high-value products, such as colours and flavours, might be produced, thus reducing dependence on the agricultural sources from which they are currently extracted.

Protein engineering techniques developed in the past decade make it possible to alter the structure of genes and thus to modify the properties of the corresponding gene products. It should be possible to apply this technology to alter not only the properties of enzymes but also those of other gene products used in food production and processing.

2.1.3 Applications using enzymes

Starch is by far the most abundant polysaccharide used in food. The starch industry is at present almost completely enzyme-based, the starch-degrading enzymes being derived from plants, animals or microbes.

The use of rennet, the common milk-clotting preparation, to produce cheese and other dairy products is by far the largest single use of enzymes by the dairy industry. Extracellular fungal proteinases can serve as rennet substitutes.

Fungi serve as sources of the lactase used to produce low-lactose milk for individuals who are lactose-intolerant. It may be possible in the future to introduce lactase-containing organisms into the milk fermentation process for the purpose of hydrolysing, within the human gut, the residual lactose in yoghurt.

The plastein reaction provides a way of using currently underused protein sources, such as leaf protein. This new technique, which involves controlled proteolysis followed by adjustment of the protein and enzyme concentrations and pH, results in resynthesis of proteins or rearrangement of peptide bonds. It has yet to be applied on a commercial scale.

The important technique of enzyme or microorganism immobilization allows continuous processing of large amounts of material. It entails the immobilization of enzymes or microorganisms on a matrix, such as membrane filters or diatomaceous earth.

2.1.4 Products used in agriculture

Biological pesticides

In recent years, interest in biological pest-control agents has increased because they have relatively narrow host ranges and thus do not affect natural predators and beneficial species. Another advantage is that pests are generally slow to develop resistance to them. Their biodegradability also means that their effect on the environment is relatively small. However, they are less stable than chemical agents, so that shelf-life is reduced and storage and handling are more costly; other disadvantages are the fact that they act more slowly on pests and that the conditions of application are more stringent.

The most commonly used biological pest-control agents include bacteria, viruses, fungi, nematodes and insects; the toxins produced by these organisms are also used. Many viruses are known to be pathogenic to insects and a dozen or more of the bacculovirus subgroup have been developed commercially. Fungal insecticides have been used successfully since the early 1900s. Other examples of organisms used as insecticides include a nematode that carries bacteria active against the black vine weevil and a wasp that is used to control a sugar-cane pest.

Biotechnology can be used to develop more efficient and potent or virulent strains, to improve the physiological tolerance of biological pest-control agents to stresses encountered in nature, and to expand the host range. The most widely used method is the cloning of the *Bacillus thuringiensis* gene for toxin production, and its transfer to another strain or species with the aim of producing a more efficient insecticide. Research is also being directed towards increasing the efficacy of the bacculoviruses through the introduction of insect-specific toxins. Many bacteria produce extracellular chitinases, i.e., enzymes that destroy chitin, a structural component of many plant pests, including fungi and insects. Chitinases have been cloned and transferred to efficient plant-colonizing bacteria.

Veterinary drugs

Biotechnological processes have been used for many years to produce veterinary vaccines and other products. Recent developments in molecular biology have made it possible to determine the structure of complex molecules of veterinary significance, such as hormones, and of the genes that control their synthesis. Thus, for example, the genes responsible for the production of porcine and bovine growth hormone have been cloned. Expression of these genes in microbial sources makes it commercially feasible to produce protein hormones, such as the somatotropins or their precursors, by the genetic modification of bacterial DNA. The production of interferon by recombinant microorganisms holds promise for alleviating the effects of viral diseases on livestock.

Genetically modified bacteria have been developed to produce products — such as vaccines and monoclonal antibodies — used in the prevention of animal diseases. While these products and their use do not constitute direct genetic modification of animals, they are noted here in order to ensure complete coverage of the broad range of possible biotechnological advances.

Rhizobia

Improvement of the nitrogen-fixing ability of cereals was an early target for those working on the genetic modification of plants. The work was initially prompted by economic considerations, but an additional consideration now is the significant environmental impact of large amounts of nitrogenous fertilizers. Nitrogen fixation occurs in many bacteria but not in eukaryotes except during symbiotic associations. Over the past 20 years, the genetics of nitrogen fixation have been elucidated. It has been demonstrated that genes coding for nitrogen fixation can be transferred from nitrogen-fixing bacteria (rhizobia) into *Escherichia coli*. This has paved the way for the transfer of nitrogen-fixation capability to plants. However, early expectations were not met and attention has therefore switched to more profitable areas of research including:

(a) the transfer of nitrogen-fixing ability, and the essential cofactors, to other species, such as *Agrobacterium* spp, which are able to infect a much wider host range than the *Rhizobium* species, which can live only in a symbiotic relationship with many legumes;
(b) improvement of the nitrogen-fixing ability of *Rhizobium* species for use with tropical legumes;
(c) improvement of the nodule-forming ability of host plants.

2.2 Plants

The development of new strains of food plants has been directed principally towards achieving improved agronomic characteristics and quality. The genetic diversity of many food crops is considerable, and this is reflected in the wide differences in chemical composition of strains of the same species.

Great progress has been made over the past 50 years in improving the agronomic characteristics of food crops using conventional breeding and selection techniques, in some instances supplemented by the use of methods to facilitate interspecific hybridization. Improvements have also been made in the degree of resistance to pests and diseases of the major crop plants, made necessary, in part, by the pressures exerted by pathogens on monocultures. Resistance to particularly devastating diseases, such as potato late blight and the cereal rusts, has been enhanced through the use of interspecific hybridization and the introduction of breeding lines selected from wild types. While the rate of progress has been slowing down because of limits imposed by the diversity of the available germ plasm, new techniques based on molecular biology should enable it to be maintained or even improved.

Attention has recently turned to the often more difficult task of improving quality and processing attributes. In many instances the biochemical basis of these attributes is poorly understood. In spite of this lack of understanding, however, there have been a number of successes, including the development of high-protein wheats and strains of oilseed rape having negligible levels of glucosinolate and erucic acid. New species have also been produced, the most successful being triticale, a fertile hybrid of wheat and rye, which combines the winter hardiness of rye with the agronomic and quality characteristics of wheat.

The breeder's task has been made easier by the incorporation into breeding programmes of new methods, including the following:

(a) Clonal propagation from organs or cells, which makes it possible to produce on a large scale genetically identical copies of superior plants. The technique is particularly useful for propagating plants for which conventional methods are inefficient, which are naturally infertile, or in which desirable traits might be lost during seed production. The technique has been used with particular success to propagate perennials such as oil palm and coffee.

(b) The production of somaclonal variants by the regeneration of plants from callus or from protoplasts, reflecting both pre-existing cellular genetic differences and culture-induced variations. Such variants provide breeders with a rapid way of increasing the genetic diversity of their breeding stock and have been used, for example, to isolate breeding lines of potato with increased resistance to disease and of brassicas with increased resistance to herbicides.

(c) The production of gametoclonal variants by regeneration of plants from cultured pollen cells or from pollen within the anther. Variation is either culture-induced or the result of recombination during meiosis. Exploited in the same way as somaclonal variation, it is particularly useful for cereal crops.

(d) Protoplast fusion, in which protoplasts from two parent plants are fused chemically, resulting in a hybrid containing a mixture of the parental genes. This technique can be used to produce new hybrid strains of plants that can be regenerated from protoplasts. For traits that are controlled by a single gene, an additional complication arises if it is desirable to ensure site-specific expression.

(e) Gene transfer, which is by far the most promising new technology. This involves the introduction of new genes into plants by means of the Ti plasmid of *Agrobacterium tumefaciens* to produce transgenic crops. The new DNA is integrated between the ends of the T-DNA of the Ti plasmid. One or more copies of T-DNA harbouring the new gene are then integrated into the plant genome at random. Transformants can be identified by selection with antibiotics, resistance to which can be included in the donor DNA. Adult plants are then reconstituted from the transfected cells, and the new genetic material is transmitted as a Mendelian trait. This process is potentially important as a means of introducing advantageous new traits into plants. For example, genes encoding substances toxic to pests can be inserted and have been demonstrated to confer resistance to predation by insects.

Ti plasmids can also be used to insert "antisense genes" in order to negate the functions of endogenous plant genes associated with an undesirable phenotype. Such genes are produced by reversing the orientation of a gene with respect to its promoter. Transcription leads to the production of RNA from the non-coding strand of the DNA, which can form a double-stranded heteroduplex with the "sense" RNA from the normal gene. These RNA heteroduplexes are not translated into protein. This strategy has already been used successfully to delay the softening of tomatoes without affecting the redness of the fruit. There are difficulties at present in inserting such constructs into monocotyledonous plants, e.g., maize, and such plants are also difficult to reconstitute from cultures.

Transposable elements, or transposons, are small DNA elements that differ from genes in that they do not occupy fixed positions in the genome, but are able to excise themselves from one site and reinsert in another, usually close to the original integration site. They can be used to construct molecular probes for retrieving the DNA that flanks them within the genome. Transposons inserting into a gene that codes for a trait of interest disrupt the gene. Recovery of the transposon and the associated DNA makes it possible to identify the gene controlling the trait, e.g., resistance to disease.

A number of food crops have been improved through modification of a single functional gene by bacterial vectors (e.g., *A. tumefaciens*), protoplast fusion, and physical methods. Such improvements include:

(a) improved resistance to specific herbicides in a number of crops, e.g., potatoes and oilseed rape;

(b) delayed softening of tomatoes during ripening;
(c) improved resistance to pests by the inclusion of genes for pesticidal substances such as *B. thuringiensis* toxin or pea lectins in a number of crops;
(d) improved resistance to viruses by the integration of genes for viral coat proteins into numerous crop plants.

In the long term it is expected that genetic modification of plants will lead to strains in which the various fractions (protein, starch, etc.) are more closely tailored to the needs of the consumer (e.g., nutritional value) or processor (e.g., functionality). It is also conceivable that crop plants might be modified so as to permit commercial production of food ingredients such as protein sweeteners.

2.3 Animals

Significant advances have already been made using conventional breeding techniques and husbandry to modify the characteristics of animals used in food production and increase the quality of their output. Changes in such diverse characteristics as maturity, fecundity and muscle distribution are observable in many domesticated animals as compared both with their wild ancestors and with domestic breeds commonly used a century or so ago.

Conventional breeding and selection techniques have enabled animal breeders to produce strains to meet producers' demands. This is especially marked in poultry, where the industry has developed early-maturing birds which, compared with the breeds used formerly, produce more eggs or can be taken to slaughter earlier. In addition, strains of cattle are available to suit high- or low-intensity agricultural systems and the demands of those wishing to produce milk and/or beef. Milk yields have been improved and the fat content of milk and flesh altered to take account of changing demands. The changes introduced into cattle and poultry by breeders have been paralleled in other commercially important species such as pigs and sheep.

In the past decade, new methods of introducing new genetic material into the germ lines of vertebrates have been developed. These techniques, coupled with the newer breeding techniques, such as cloning, have increased the speed with which improved strains can be introduced and offer significant potential for advances in food production. New methods of gene insertion include microinjection of DNA into the pronucleus of the fertilized egg, the creation of recombinant retroviruses which can infect embryos and insert their DNAs into the genome of the host, and embryonic stem (ES) cell systems.

Microinjection is conceptually the simplest and by far the most widely used technique, though it cannot be used in fowl. Typically, the technique entails the physical insertion of linear DNA molecules into the pronucleus of the one-celled zygote. After the zygotes develop into viable progeny, they can be screened by a variety of techniques for incorporation of the new genetic material. This process is associated with high levels of expression of donor genes. The tissue distribution and developmental regulation of foreign gene expression depends on the elements included within the donor sequences. Such elements are commonly referred to as promoters or promoter/enhancer complexes. By means of recombinant DNA technology, a promoter from one gene can be grafted onto another to target expression of the new coding element to the tissue of choice. As an example, growth hormone expression can be elicited from liver by inserting growth hormone genes equipped with a liver-specific promoter (e.g., albumin). Another advantageous feature of microinjection is that there is no apparent limit to the size of the DNA molecule that can be inserted.

Retrovirus-mediated gene transfer involves the replacement of most of the genes of these RNA viruses by new genetic material. The constructs are introduced into "packaging cell lines", and these generate packaging proteins that allow formation of viral units which are then harvested from culture media and used to infect embryos. Once in the embryo, the recombinant DNA is no longer infectious because it lacks the relevant genes for packaging. The usefulness of retrovirus-mediated gene transfer is limited because a regulatory element (the long terminal repeat) required for completion of the gene transfer cycle inhibits expression of transferred genes. In addition, there is a limit to the amount of new DNA that can be packaged for transfer. Because a larger number of avian viruses have been characterized, this technique is more likely to be applied to fowl than to mammals.

ES-mediated gene transfer is made possible by the development of ES cell cultures. ES cells are derived from preimplantation embryos and can be maintained in culture for prolonged periods and subjected to genetic modification. When incorporated into the embryonic lineage of a developing embryo, they differentiate into all cell types including germ cells. The resulting animal consists of both cells derived from the ES line and those descended from the recipient blastocyst. Progeny from ES-derived germ cells (usually the sperm of male founder animals) carry the trait in all their cells, including eggs or sperm.

An important limitation in all methods of gene transfer is the identification of strategies for introducing advantageous genetic changes into mammals. Introduction of growth-promoting genes into fish has been highly successful in enhancing somatic growth, and this technique should result in significant future advances in aquaculture. However, such genes do not have such a pronounced effect in swine. Despite these difficulties, there have been several important applications of gene transfer in large animals, including growth enhancement and the introduction of drug-resistance genes. Other applications include enhancement of the immune response, insertion of genes whose products block infection by viruses, alteration of the fat and/or cholesterol content of meat, and modification of intermediary metabolism so as to alter the pattern of protein or fat production. It should also be possible to alter the nutritional profile of milk by the insertion of new genes. Most practitioners of this technology estimate that at least some of these goals will be achieved within the next ten years.

Although not directly related to food production, a more immediate and profoundly important application of transgenic animals involves the insertion into livestock of genes for regulatory signals that allow release of proteins into their milk. This approach is useful for producing complex medicinal compounds, such as clotting factors, while avoiding the risk of contamination by human pathogens, such as hepatitis or human immunodeficiency virus. Experimental systems have already shown that biologically active human clotting factors can be released into the milk of transgenic animals by linking the relevant genes to promoters from milk protein genes.

2.4 Food analysis

The application of biotechnology has resulted in the development of rapid and sensitive analytical methods which may have many applications in food analysis. These methods, which include the use of DNA probes and immunoassay methods, are expected to prove useful in improving the detection of food contaminants.

Knowledge of the DNA structure of target organisms has allowed characteristic gene sequences to be identified and, from these, DNA probes have been constructed, specific to the target gene sequence. Such DNA probes can be used both to differentiate between strains of the same species of organism and to determine whether particular microorganisms are capable of producing specific toxins.

Immunoassay methods, involving the use of monoclonal antibodies, provide simple diagnostic tools for both microbial and nonmicrobial contamination and can be marketed in kit form. These can then be used for a number of purposes, based, as for DNA probes, on isotopic or nonisotopic detection methods. Recent advances in nonisotopic methods have led to the development of highly sensitive and rapid detection systems.

3. Safety assessment of foods derived from microorganisms generated by biotechnology

3.1 Introduction

Because microorganisms have been used for centuries to produce foods and food ingredients, considerable experience has been gained regarding the factors that need to be taken into account in ensuring that such substances are safe. The new molecular techniques for introducing and modifying heritable traits in microorganisms have greatly expanded the pool of genetic traits available for strain improvement. These new methods permit specific, well-defined sequences, including genes from plants and animals, to be introduced rapidly into a variety of microbes.

The Consultation considered the scientific principles judged to be important in ensuring that foods and food ingredients produced from microorganisms are safe for the intended use. The Consultation was aware of the great diversity of the products of biotechnology, including foods and food ingredients. It recognized that the nature of these substances calls for the evaluation of a number of factors, which may vary with respect both to the food or food ingredient in question and to the proposed use of the product. The chemical nature of the substance to be produced and the amount that will be used in food are critical factors to be taken into account in a food safety assessment. For microbes, it is important to consider whether the organism itself will become a component of food, as is the case for starter cultures and probiotics, or whether the substance to be added to food will be isolated from the production culture by extraction either from cellular material or from the supernatant liquor.

The substances that are the subject of this section can be classified as follows:

1. Microbes used as sources of food substances produced by fermentation, such as amino acids, organic acids, flavours, thickeners, antioxidants, preservatives, and enzymes used in food

processing. They are also used for the production of food ingredients (e.g., single-cell protein). In these applications, viable organisms are not intended to be part of the finished food.
2. Microbes used as essential constituents of food in order to produce characteristic effects, such as acid and flavour (e.g., dairy starter cultures), or microbes intended for use as probiotics. These applications involve the addition of living microorganisms to food.

The Consultation recognized the importance of other substances produced by microbes, such as pharmaceuticals and biological control agents, which may appear as residues in food, but did not consider that such substances came within its terms of reference.

3.2 Issues to be considered in safety assessment

The Consultation identified several potential hazards that could result from the use of genetically modified microorganisms to produce food or food ingredients, including:

(a) microbes that produce toxic substances or are pathogenic;
(b) introduced genetic material (e.g., vectors) that encodes harmful substances or that can be transferred to other microbes in which it would encode harmful traits;
(c) the transfer to pathogenic microbes of marker genes that encode resistance to clinically important antibiotics, rendering the recipient organism refractory to antibiotic therapy;
(d) the production of unexpected products (or increased levels of normal products) of genes, effects on multiple genes (pleiotropy), or secondary effects that result in harmful substances in the finished food;
(e) potential adverse consequences of modified microbes in the human gastrointestinal tract;
(f) possible adverse effects (e.g., allergenicity) of new or altered proteins;
(g) possible adverse nutritional changes in food;
(h) possible adverse effects of changes in the structure and/or function of gene products.

These potential hazards are not new and can be controlled adequately using sound microbiological principles. Indeed, the techniques of molecular biology greatly enhance scientists' ability to address these concerns and to ensure safety.

Microorganisms traditionally used in food can generally be grouped in categories according to previous experience in their use and known risks, and different levels of concern can be assigned reflecting the available scientific information.

3.2.1 General considerations

The safety from both a toxicological and nutritional point of view of a food or food ingredient produced by a microbe during fermentation will depend on all the steps of the process, namely: (*a*) the source organism, its characteristics, and how it has been developed to perform its intended function; (*b*) the fermentation process, including the substrates and other growth materials and the conditions of growth; (*c*) the isolation procedures and all the steps in the purification process, when appropriate; and (*d*) the final standardization of the product.

Each step of the process must be carefully evaluated. An evaluation must also be conducted when any change is made in the microbe or the production process. Current good manufacturing practices must be fundamental to any production process.

While all the factors mentioned above must be taken into account in assessing safety, this section focuses on the production microorganism *per se* and the factors that generally need to be evaluated to determine whether the food or food ingredient produced will be free of unsafe impurities from the production source. In some instances, factors other than those discussed below will need to be included in the assessment, while in other cases some of the factors discussed here will not be relevant.

In addition, scientific advances will show that new aspects need to be considered and lead to new methods of assessing safety. It is therefore recommended that, rather than laying down specific requirements, the approach to assessing the safety of microorganisms should be to draw up broad terms of reference based on sound scientific data while retaining sufficient flexibility to accommodate scientific advances. Such an approach has already been applied successfully by the Joint FAO/WHO Expert Committee on Food Additives in evaluating the safety of food additives, including those produced by microorganisms (*3*).

The safety assessment paradigm presented in section 3.3 places emphasis on the use of microbiological, molecular, and chemical data in evaluating the safety of foods and food ingredients produced by microbes, and the use of such data on a case-by-case basis to determine the testing necessary. The application of the principles discussed to microbes that have established safety records and the demonstration that the substance of interest, or in certain cases the microbe, is sufficiently similar to its traditional counterpart should normally provide a reasonable degree of certainty of the safety of the food, with minimal supplementary testing.

Human health must be a critical element of any safety assessment, whether related to food or the environment. Measures should be taken to support and ensure the use of the same scientific rationale in both aspects of the safety assessment. The Consultation did not focus on environmental issues but noted that a document published by the Organisation for Economic Co-operation and Development deals with this issue (4).

3.2.2 Specific safety assessment considerations

Safety assessment for human health should cover potential pathogenicity, toxigenicity, and nutritional considerations.

Characterization of the genetic modification

The new methods of molecular biology permit essentially any gene to be transferred to any recipient organism, regardless of natural biological barriers. Both biological (e.g., plasmid or phage vectors) and physical (e.g., electroporation, microprojectile) methods can be used to introduce DNA fragments encoding new heritable traits into recipient (production) organisms. It is essential that genes and additional DNA introduced into the production organism be characterized so as to ensure that the introduced DNA does not encode any substance that would be harmful. In addition, data should be provided demonstrating that the final construct is stably maintained. Analysis of DNA sequence data for open reading frames and regulatory signals, and confirmation of the expected genetic construct by appropriate methods, such as sequencing, restriction mapping and hybridization techniques, should be standard practice. Other physical and chemical techniques (e.g., analysis of transcription products and protein patterns) may be useful in confirming that the expected modifications have been achieved. All the steps leading to the final construct should be evaluated and confirmed by appropriate techniques.

All organisms used to construct the production organism (i.e., organisms that donate genetic material or are used as intermediate donors) should also be well characterized.

Origin and identity of the host organism

The use of host organisms that have been shown to be safe for use in food is preferred; the safety of microbes that do not have such a history must be established by means of appropriate tests.

The taxonomic identity of the production (host or recipient) organism should be established both genotypically and phenotypically. If the host organism has been obtained from a research collection, its taxonomic identity and traits should be confirmed; if obtained from other sources (e.g., the natural environment), its taxonomic identity should be established by comparison with reference cultures. Although scientific advances may occasionally alter the accepted taxonomic classification of a culture, taxonomic information is an important aid in identifying factors critical to safety, such as potential toxins (including clinically useful antibiotics), virulence factors, and other impurities, that should be evaluated. The new methods of molecular biology permit organisms to be more easily identified and placed in their proper taxonomic niche.

The addition, deletion, or modification of one or several traits in a production organism will rarely change the fundamental taxon to which that organism belongs.

The pedigree of the production organism should be evaluated, including all known genetic modifications, whether effected by new or traditional methods.

Vectors

Vectors, such as plasmids, phages, or cosmids, are the vehicles used for transferring genetic material to recipient hosts. Since the final construct must be suitable for use in food, the vectors must be well characterized so as to ensure that no harmful substances are produced. The vectors already used in food biotechnological applications have been completely sequenced and shown not to encode harmful substances. Examples include pBR322, which has been modified to contain the bovine prerennin gene for production in *E. coli* K12, and pUB110, which has been modified to contain the α-amylase gene obtained from several donors for production in *Bacillus subtilis*.

Several investigations are developing "food-grade vectors" intended to be safe for use in food processing (5–7). Such vectors generally have the following characteristics:

1. They are derived from organisms recognized to have a history of safe use in food.

2. Any heterologous genetic material introduced is well characterized; generally, such genetic material should be the smallest fragment necessary to express the desired trait.
3. They do not contain selectable marker genes that encode resistance to clinically useful antibiotics, if the organism is to be used in food (see below for a discussion of marker genes). If the organism is the source of an ingredient, the finished ingredient preparation should not contain viable cells or biologically active DNA that encodes resistance to antibiotics.
4. They are modified so as to minimize the transfer of traits to other microorganisms.
5. Mobile genetic elements are modified so as to minimize the instability of the construct and reduce the possibility of transfer to other microbes.

The products of newly added genes should be evaluated in order to establish that the substance is safe for the intended use in food; the evaluation should be based on conventional food safety considerations, such as chemical structure and dietary exposure.

Selectable markers

One of the most useful methods of isolating and selecting new varieties that have received an intended genetic trait, chromosomally or extrachromosomally, is based on the use of genes that encode resistance to antibiotics such as ampicillin, tetracycline, and kanamycin. The transfer of such genes to pathogenic microbes can render the organism refractory to clinical therapy. For this reason, functional marker genes should not be present in the final production microorganism when the viable organism is to be used in food (e.g., starter cultures). Research is being conducted to develop alternative markers (5). The safety of such gene products must be evaluated on a case-by-case basis.

The use of antibiotic-resistance genes as selectable markers in microbes that are not intended to exist as viable cells in a finished food substance raises a somewhat different question. The finished substance should be examined, using properly controlled assays, to ensure that it is free of biologically active (transformable) DNA that could encode resistance to antibiotics. Techniques such as gel electrophoresis coupled with DNA hybridization can also be used to determine whether any segments of DNA occur in the finished substance that would be of sufficient size to encode the antibiotic-resistance gene.

The final construct

The new methods of molecular biology provide better means of regulating the expression of desired genes. This can be achieved through the use of improved regulatory signal sequences (e.g., promoters, enhancers, ribosome-binding sites), gene amplification (duplications), multicopy plasmids and other means. These techniques are not a cause of concern in and of themselves, but the increased levels of the products of genes under such control should be considered in the light of conventional food safety considerations.

Antisense mutagenesis provides a relatively new means of modifying the level of gene expression. In this technique, a segment of the gene of interest is inserted in reverse orientation into the host organism. The antisense RNA produced interferes with expression of the target gene, reducing the amount of protein produced.

Concern has been expressed that antisense modulation of gene expression may also reduce the expression of genes "downstream", thus causing secondary effects. In addition, insertion of genetic material in a region of the genome that controls the expression of other genes can result in pleiotropic effects. The number of examples of genes that exert global responses and control regulons such as heat-shock proteins is continually increasing.

Genetic modifications, whether effected by new or traditional methods, may give rise to pleiotropic effects; however, while the possibility of such effects must always be remembered, the new methods do not increase the likelihood that they will be harmful when they do occur. Careful consideration of the target sequence involved will permit predictions of secondary effects on the cell of both new and traditional means of genetic modification.

Pathogenicity and toxigenicity

Microorganisms intended for use in food processing should be derived from organisms that are known, or have been shown by appropriate tests in animals, to be free of traits that confer pathogenicity. Pathogenicity in microorganisms is the result of a combination of traits (e.g., adhesion, invasion, cytotoxicity). For this reason, the addition of one or a number of well-characterized genes not coding for pathogenic traits to a production organism that was previously established to be nonpathogenic, will not convert that organism into a pathogenic strain (8).

A number of factors must be considered in connection with toxigenicity, including known microbial toxins such as enterotoxins, endotoxins, and mycotoxin-like substances, and antimicrobial substances. The potential production of such substances can be assessed on the basis of the nature of the production organism, how it has been genetically modified (including potential pleiotropic effects), how it is grown for its intended purpose, and how the finished substance is purified. Microbial enterotoxins are well known in certain species of bacteria involved in outbreaks of foodborne illness. These toxins are active by the enteral route, and the possible presence of such toxins should be evaluated, on the basis of the identity of the organisms used to construct the production organism. Introduced DNA can be examined for the absence of DNA sequences of the size and pattern of known toxin sequences, and in case of doubt, animal and *in vitro* tests can be used to confirm that toxins are absent. Appropriate methods, such as the polymerase chain reaction and the use of DNA hybridization probes for toxin genes, can confirm the absence of sequences that could encode such genes. A general acute feeding test in an animal that exhibits a diarrhoeal response can be useful in confirming that biologically active levels of unexpected acute toxins are not present in the finished food substance.

Endotoxins are substances associated with the cellular envelope of Gram-negative bacteria and can have important implications in medical applications, e.g., producing shock if accidentally given with drugs administered intravenously. With food substances that are ingested, the concern is not as great because the toxins are not active in the human gut at the normal background levels found in drinking-water, milk, and other foods. If endotoxins are believed to be present in the final food substance obtained from a Gram-negative production organism, appropriate tests may be performed on the substance to ensure that the levels do not significantly exceed the normal background levels.

Microorganisms are known to produce under certain growth conditions a number of other substances that are toxic to various organisms, including humans in some cases. These include mycotoxins, antibacterial substances (e.g., bacteriocins from lactic acid bacteria), and antibiotics (e.g., bacitracin from *Bacillus* spp).

Fungi (e.g., *Aspergillus* spp) are well known to be capable of producing toxins; this production is highly dependent on the strain of microorganism and the conditions under which it is grown. These issues have been discussed by the Joint FAO/WHO Expert Committee on Food Additives (9); however, it has been pointed out that these substances are not known to be produced under conditions of current good manufacturing practice. The Consultation nevertheless took the view that vigilance and careful attention to good manufacturing practice are necessary.

The use of the new molecular methods does not raise new concerns with regard to the potential for the production of such toxins, provided that careful consideration is given to the identity of the microbe, its genetic make-up and physiology, the growth conditions, and the final food or food ingredient. Where questions arise about the safety of production organisms because of possible mycotoxin production, appropriate chemical assays and toxicological testing are recommended (9, 10).

Bacteriocins are not known to be toxic to humans; instead, they provide the producing organism with a means of eliminating other microbes in the natural environment. There is considerable interest in developing bacteriocins for use in food processing, and one, nisin, is currently used in cheese-making. The safety of these substances can be evaluated on the basis of conventional considerations, such as identity and dietary exposure, and in the light of the microbiological principles discussed here.

The use of microbes as probiotics for the colonization of the intestinal tract of humans or animals calls for additional factors (e.g., survival, growth, colonizing ability, and transfer of genes to other intestinal microbes) to be taken into account. The production of physiologically active ingredients in the food or gastrointestinal tract requires special consideration.

The question of the possible increased risk of allergies from ingestion of new or altered proteins is frequently raised. This question is difficult to address scientifically because only certain proteins and glycoproteins elicit such reactions and generally only a relatively small proportion of the population is sensitive. In addition, there are, as yet, no reliable tests from which allergenicity can be predicted. Proteins used in food and food processing are largely digested in the gastrointestinal tract. As a result, they do not arouse the same degree of concern as proteins introduced into the body by injection, e.g., drugs and biologicals. Variation in chemical structure is common in classes of proteins traditionally used in food processing due to differences in the degree of glycosylation and even in amino-acid sequence, and this suggests that such variation is not generally of particular concern. However, food allergies do exist, and it is known that a small fraction of ingested protein is absorbed into the body (11). If the gene for a protein has been obtained from a source (e.g., wheat or peanuts) known to be highly allergic, the possibility that the cloned protein may induce sensitivity should be considered. Comparison of gene sequences with data on known allergens may also become increasingly useful as the information on such proteins increases.

The possibility of any significant adverse changes in the nutritional composition of a food as a consequence of microbial action should be evaluated by comparison with the nutritional composition of the conventional counterpart.

In the overall assessment of genetically modified microorganisms, and the biological containment of microorganisms to be used for fermented foods and as probiotics, it is important to take into account the characteristics of the microbial strains that determine their survival, growth and colonizing potential in the gut, including the capability to undergo transformation, transduction and conjugation, and to exchange plasmids and phages. A general principle is that design should be directed towards minimizing intrinsic traits in microbes that allow them to transfer genetic information to other organisms. With regard to vector design, special safety cassettes in such microbes might be of interest in reducing the possibility of the transfer, e.g., of drug-resistance genes, to pathogenic bacteria in the intestinal tract.

Chemical identity of the gene product

In the application of the new methods of molecular biology, circumstances may arise in which the chemical structure and functional activity of a substance produced in a recombinant-DNA-derived microbe will differ from those of the traditional substance. Molecular biology also provides new and more precise ways of predicting and determining structure.

With some minor exceptions, the nucleic acid sequence of a gene will produce the same polypeptide sequence regardless of the genetic background in which it is expressed. Differences in codon preference in various organisms may necessitate minor changes in the inserted DNA sequence in order to accommodate the host system, but the amino-acid sequence can generally be predicted from the DNA sequence of the gene. Other structural changes may occur in proteins as a consequence of post-translational processing, e.g., glycosylation or methylation. As a consequence, there is considerable variation in both the structure and biological function of many substances that have been used in food. For example, the starch-hydrolysing enzymes vary considerably because of differences both in post-translational processing and in primary sequence and function (*12*). α-Amylases from *B. subtilis* that are identical except for one amino acid (serine or tyrosine) vary in their ability to hydrolyse maltotriose to glucose and maltose (*13*). Such variation demonstrates that many food substances should be considered in terms of the category to which they belong, rather than as individual entities. The modifications introduced by means of the new biotechnology should be considered in the light of the degree to which the properties of the new substance differ from the range of properties of known substances in its category. In most cases, minor differences in structure and function will not give rise to safety problems.

3.3 Safety assessment paradigm

The new techniques of molecular biology provide improved and more precise means not only of making genetic modifications but also of confirming that the expected modification has indeed occurred. To the extent possible, the sequence of introduced genetic material, hybridization data, and other physical and chemical data should be used to determine whether the substance produced by the introduced DNA is as predicted, allowing for minor variations. When this is the case and the substance in question is sufficiently similar, chemically, structurally and biologically, to substances that have a history of safe use in food, the remaining safety questions will centre on impurities derived from the source organism and other processing aids and, for food substances produced by fermentation, the purification process. Provided that any processing materials are of approved food grade and the production organism has a history of safe use, only minimal animal testing should generally be needed to ensure that unexpected toxins are not present. This is consistent with proposals for evaluating microorganisms and substances produced by microbes based on a combination of the case-by-case and decision-tree-type approaches (*11, 14*).

Every process for the preparation of a food or food ingredient results in a mixture of chemicals, the composition of which will vary, depending on a number of factors. An organic acid or amino acid will have a specific structure, but the impurity profile of commercial preparations of these substances varies, depending on how they are produced. Food substances, such as enzyme preparations, are complex mixtures of active substances and impurities. The active substance, e.g., the amylase, lactase or lipase protein, can therefore be viewed, as already suggested, as belonging to a class of substances within which considerable variation in structure and function occurs. For substances that have commonly been used in food processing and are used in food at low levels of dietary exposure, such variation is relatively trivial, and the focus of the safety assessment should be on the nature of the production organism, how it has been modified (including characterization of inserted DNA), how the substance is obtained by fermentation and purified, and its intended use in food.

Generally, food additives that have a history of safe use in food but which are produced by new technology will not require extensive toxicological evaluation or the establishment of new acceptable daily intakes (ADIs). Provided that the substances produced are as predicted (allowing for minor differences in structure and function) and do not contain impurities of an uncertain nature, they may be considered to be sufficiently similar or equivalent to their conventional counterparts. Others should be evaluated on a case-specific basis.

It is imperative that the development of microbes for use in food processing should be in accordance with current good laboratory practices and that manufacturers should establish and follow good manufacturing practices in the production of foods and food ingredients.

3.4 Summary

The new techniques of molecular biology open up great possibilities for important developments in the global food supply. At the same time, these techniques provide sound scientific methods for addressing questions of the safety of the foods and food ingredients produced by their use. In many applications of the new biotechnology in which improved microorganisms are used in food processing, a careful, rigorous analysis of the microbiological, molecular, chemical and toxicological parameters will provide a sound basis of safety when the food substance is produced in accordance with current good manufacturing practice.

4. Safety assessment of foods derived from plants generated by biotechnology

4.1 Introduction

Many plants used for food contain natural toxins at levels that can vary widely between different strains of the same species. Levels may also vary in plants of the same strain as a result of the influence of external factors such as climate, season and location. In a very small number of instances, new varieties bred by conventional means for improved agronomic and processing characteristics have been found to contain unacceptably high toxin levels. In addition to the well known example of the potato variety, Lenape, which was developed for good processing characteristics, but which was found to contain unacceptably high levels of solanine, unacceptable levels of cucurbatin in vegetable squash strains and psoralin in celery strains have also been found.

In most countries, the safety of traditional foods derived from plants is not regularly assessed on the grounds, in part, that foods consumed for generations should be safe for consumption in a normal diet. With the development of techniques that permit rapid and marked genetic modification of crop plant cultivars, sometimes greater than what would be expected in natural gene transfer, there is a need to formalize the process of food safety assessment for crop plants.

The possible hazards are both toxicological and nutritional in nature, and safety assessment must address both types. To ensure that the final product is safe for consumption, information is required on both the technology employed to modify the plant and the effects of the genetic modification. This should allow the correct questions to be posed to establish safety in use. It is important that the significance of any hazard is considered in relation to those associated with existing food sources and those that might be introduced by the genetic manipulation of crop plants in traditional plant breeding.

Although this assessment of food safety has been prompted by the new techniques of genetic modification via recombinant DNA technology, the possible hazards are also associated with other forms of genetic modification in plants.

4.2 Issues to be considered in safety assessment

4.2.1 General considerations

A variety of techniques can be used to transfer foreign genes into plant cells. Whatever method is used, only a small amount of DNA is integrated into the plant genome (usually 5–25 kilobases). This DNA is usually a component of, or a complete bacterial plasmid, designed and constructed in a specific manner. The precision of recombinant DNA technology allows such vectors to be well characterized. The complete nucleotide sequence of the DNA region to be transferred to plant cells is usually known or can be derived from the component fragments, and the nature of all encoded functions is understood.

Transferred DNA is integrated into the genome of the recipient plant essentially in a random manner, since genetic engineering does not yet allow the position of insertion to be predetermined. Furthermore, there is no control over the number of integration events, or whether the DNA transferred is complete, truncated or rearranged. However, although integration cannot be controlled, its outcome can be determined via DNA analysis of the transgenic plants. Once regular expression and inheritance patterns are observed, they are maintained in a stable manner, in the same way as characters introduced by traditional plant breeding.

Inserted genes

The DNA from all living organisms is structurally similar. For this reason, the presence of transferred DNA in produce in itself poses no health risk to consumers. One issue often raised is that of the transfer of genes from plants to bacteria. However, there is no well documented evidence of this or known biological mechanism by which it could occur. Nevertheless, it cannot be absolutely guaranteed that DNA will not be released from food in the digestive tract and taken up and integrated by the gut microflora. If this does happen, it would have to be as the result of a large number of discrete steps, each of which would only be expected to occur extremely infrequently. The possible transfer of antibiotic-resistance genes used as selectable markers for plant transformation is of particular concern. However, the risk involved is considered to be insignificant compared with that associated with other mechanisms by which microbes can become resistant to antibiotics.

Expression products

The expression of the transferred genes is under the precise control of specific regulatory signals, which determine where, when and to what extent the gene product is made. Following the selection of transgenic plants, such expression patterns are highly stable and predictable. In many instances the transferred genes need not be expressed in the harvested produce of crop plants. When they are so expressed, an informed judgement can be made as to their safety, since the nature of the expression products is usually well known.

Pleiotropic effects

Some of the genes transferred to crop plants will encode enzymes that catalyse specific steps in biochemical pathways. Such gene expression may result in a depletion of the enzymatic substrate and a concurrent accumulation of the enzymatic product. Since the biochemical basis of the expression products derived from the transferred genes is usually well understood, these pleiotropic effects will be largely predictable and can therefore be evaluated. Pleiotropic effects can also occur in association with gene transfer by the traditional methods of plant breeding.

Secondary effects of gene insertion

The transfer of a gene specifying a higher level of activity of one enzyme in a biochemical pathway could result in an increased flow of metabolites through the downstream steps of that pathway. These changes are largely predictable and can therefore be evaluated. The phenotypic changes resulting from these potential secondary effects are the same as those possible in traditional plant breeding, especially following wide hybridization. The progeny resulting from such wide crosses could conceivably exhibit identical secondary effects resulting from the complementation of genes specifying enzymes with different levels of activity along a biochemical pathway.

Secondary effects of gene disruption

The use of antisense technology allows the expression of a specific gene to be partially or completely shut down. This will also decrease or eliminate the flow of metabolites through downstream biochemical pathways. There may also be an accumulation of the substrate for the targeted enzyme which may, in some instances, result in the shunting of metabolic flow into other pathways. These possible effects are largely predictable, which allows any specific hazards to be identified and evaluated. They do not constitute a new health hazard since identical situations can arise as a consequence of mutagenesis in conventional plant breeding.

Insertional mutagenesis

Transferred genes are generally considered to be inserted randomly into the genome of recipient plants. Such insertion events may disrupt or modify the usual expression of genes in the plants. Thus inactivation may result from insertions into the coding regions of existing genes, thereby disrupting their normal functioning. Similarly, activation may result from insertions into the regulatory regions of existing genes, thereby modifying the manner in which they are expressed. However, the vast majority of DNA in plant genomes is noncoding or repetitive, and insertions into these areas would be expected to have no significant effect on the plant phenotype. Insertional mutagenesis events should therefore be rare and, if they do occur, inactivation would be expected to predominate. They do not constitute a new health hazard since similar situations arise in traditional plant breeding as a consequence of standard mutagenesis, natural genetic recombination, chromosomal rearrangements (translocations and inversions) and the activity of transposable elements. The main potential hazard of insertional mutagenesis from the point of view of food safety is the activation of silent genes in the harvested produce of the crop, especially genes that may direct the biosynthesis of toxic secondary compounds. The potential toxicants for which screening is necessary are those present in the nonharvested portions of the crop plant and in closely related species.

In some instances each independent transformation event may give a different response, even when identical vectors and the same plant genotype are used. This is particularly true of potential hazards associated with random insertional events. The screening of large numbers of independent transgenic plants for identifiable hazards will substantially reduce the chances of releasing a cultivar yielding a harvested food containing undesirable constituents.

The precision with which the nature and expression of the DNA transferred can be controlled by means of recombinant DNA technology means that, as compared with the traditional methods of genetics and breeding, this technique provides greater assurance that the desired outcome will be achieved. For this reason, it is important that food safety assessments proposed for transgenic plants are also applied to new cultivars of crop plants derived from traditional genetic modifications. The same issues will also arise in relation to genetically modified plants used in feed for animals intended for human consumption.

The use of modern technologies may result in products that differ both toxicologically and nutritionally from the traditional ones. However, it should be noted that the presence of toxic materials, or changes in the nutrient content or the bioavailability of nutrients may not in themselves preclude marketing of the product.

The *toxicological* changes in products that may result from the use of modern technologies include:

(a) the presence or increased content of natural toxicants;
(b) the presence of new expressed toxic materials resulting from genetic modifications (e.g., biopesticides);
(c) development of allergenicity;
(d) accumulation of toxicants or microbial contaminants derived from the environment;
(e) changes in the availability of toxins as a result of processing.

The *nutritional* changes include:

(a) modification of major nutrients, micronutrients or antinutrients in the food;
(b) changes in the bioavailability of macro- and/or micronutrients;
(c) changes in nutritional components as a result of processing.

It is difficult to perform toxicological assessments on whole foods by the traditional methods of toxicity testing. One difficulty is the absence of good animal models for some of the end-points of interest for such foods. Testing can also be difficult because of the inability to amplify adequately the exposure to potential toxicants and to account for confounding factors, such as nutrient imbalance. For these reasons, a safety assessment based on the use of molecular, biological and chemical information should initially be carried out.

The technology, in part, determines both the type and quantity of information required to make an assessment. For example, modifications that result from the transfer of a known gene from one variety to another by homologous recombination will be of less concern than the generation of a new variety with a novel trait, e.g., pest resistance or hardiness, through unknown mechanisms and by unspecified alterations in the host genome.

The product to be assessed is the food plant and, more specifically, those parts of it that will be consumed. The standard used for purposes of comparison must be the traditional plant or food. As with all food, the product that is marketed should be evaluated, taking processing into account. Estimates of level of exposure to new foods and expression products in foods constitute a critical component of any safety assessment.

4.2.2 Specific safety assessment considerations

The components of the safety assessment include:

(*a*) characterization of the host, donor/vector and modified organisms;
(*b*) characterization of the precision of the process and possibility of pleiotropic or secondary effects;
(*c*) characterization of novel expressed material;
(*d*) characterization of the modified food for wholesomeness.

While it is the final product that must be tested for safety, some data generated in the initial stages should be acceptable in the assessment, in particular data on the method of modification, the genetic stability of the transformed plant and its molecular biology. For transgenic plants, some of these data will be collected during the initial small-scale field trials, the majority being collected during scaled-up field trials as the selected lines for possible release are gradually identified.

Characterization of the host, donor/vector and modified organisms

The starting point will be the characterization of the host and donor organisms. Such information may include *inter alia*: (i) known toxins produced by the plant itself or other organisms commonly associated with it; (ii) relationship to toxin producers belonging to the same genus; (iii) use as food; and (iv) medicinal use.

Many traditional food plants either contain, or have close relatives that express, natural toxicants. Detailed information on the presence, ability to express, and possible exposure to these materials will be required.

The modified plant should be assessed with respect to stability of gene expression and inheritance and its general phenotype. Modifications, such as resistance to herbicides and heavy metals, should also be evaluated on the basis of metabolism, possible mechanism and residue analysis.

Characterization of the precision of the process and possibility of pleiotropic or secondary effects

Sufficient information on the method of modification used should be obtained to enable both the safety of the process and the possibility of pleiotropic and secondary effects of the process to be assessed.

Information on all aspects of the genetic manipulation should be obtained to a degree consistent with the available technology. All available information on food uses should be provided.

In plant transformation, it will often be possible to restrict the expression of the inserted genes to specific plant tissues. Detailed information on where, when and to what magnitude the gene is expressed will then be necessary.

Characterization of novel expressed material

Newly expressed material, whether introduced or modified native material, proteinaceous or nonproteinaceous, should be characterized and appropriate toxicity tests performed where relevant. If the material constitutes a significant proportion of the edible portion of the plant, it may be necessary to evaluate the nutritional characteristics of the whole food instead of the constituent *per se*. Where expressed material is expected to have a physiological or toxicological effect, it should be evaluated in a fashion similar to that adopted for the parent material.

Characterization of the modified food for wholesomeness

It is generally accepted that the composition of traditional food cultivars varies widely as a result of both genetic and environmental effects. A comparison should initially be made with the parent (host) plant and, if necessary, with other cultivars of the host type. Two possible results are envisaged: (*a*) modifications in the content of traditional constituents; and (*b*) the appearance of new constituents. The latter should trigger more detailed studies on their characterization, and possibly also requirements for toxicity data either on the new constituents or on the modified plant food. The degree of detail required in an assessment of safety based on theoretical grounds or chemical analyses is process-specific and will depend on the nature of the genetic change, the genetic technology used to make the change and the particular crop plant in question.

Since many of the projected novel plant foods involve modifications in proteinaceous food components, and possibly the formation of new components that are potentially allergenic in character, this issue should also be addressed.

With many novel plant foods, nutritional concerns will be the major issue, especially where major nutrients are modified or eliminated. In addition, the possible loss or reduced bioavailability of micronutrients and increases in antinutritional factors should also be addressed.

Detailed compositional analysis of the edible portion of the plant will be required. In the light of the importance of the food and the specific nutrient in the national diet, such analysis may include composition, levels of true protein and nonproteinaceous material, amino acid profile, composition of carbohydrate fraction, qualitative and quantitative composition of total lipids, inorganic components, and vitamins, naturally occurring or adventitious antinutritional factors, and storage stability from the point of view of nutrient degradation. Where the food is a major constituent of a diet, bioavailability studies may also be necessary.

4.3 Safety assessment paradigm

Traditionally, new varieties of food plants have not been subjected to extensive chemical, toxicological or nutritional assessment. Exceptions are foods intended for specific consumer groups (e.g., infants, diabetics), where the food may form a substantial portion of the diet. The safety assessment of novel foods calls for both a new paradigm and a new framework of assessment. To date, all assessments of novel foods have been made on a case-by-case basis. Suggested model frameworks have included decision-tree analysis, full toxicological work-up and limited assessments based on background information.

A new paradigm for safety evaluation, with the emphasis on molecular, biological and chemical data and the use of these data to determine the need for appropriate toxicity tests, is recommended. This new paradigm for safety assessment should be applied to all forms of genetic modification in plants, including traditional plant breeding.

The components of the safety assessment process are essentially suggestions as to the information that may be requested, rather than requirements specifying what should be provided. The detail required will depend on the nature of the genetic modification.

If the assessment based on the molecular, biological and chemical data does not allow an informed judgement on safety, conventional assessment procedures based on animal studies will be needed (*3*).

4.3.1 Animal studies

The need for toxicity testing will be determined in part by the nature of the modified plant food. Molecular, biological and chemical analyses should always be conducted before the need for animal testing is assessed. When the assessment of genetic or compositional change does not provide a satisfactory basis for safety evaluation, it may be necessary to test the whole food by means of appropriate animal tests. The nature and extent of such testing must be carefully assessed in relation to the need to provide additional assurance of safety.

For new or unexpected constituents with no history of safe use, appropriate toxicological testing must be performed to ensure the safety of the food under the conditions of use. Limited testing of the food in feeding studies may sometimes be sufficient. Where newly expressed material has known physiological or toxicological effects, it may be appropriate to conduct mutagenicity studies on the purified material. It will usually be possible to conduct these tests on the constituent itself. Occasionally, however, it will be necessary to test the safety of the whole food, e.g., when the isolation or synthesis of individual constituents is not possible. Toxicological tests on the whole food must be carefully designed and controlled, and appropriate controls, including the traditional food, should be used. Careful attention should be given to understanding and controlling confounding factors.

4.3.2 Human data

It is becoming increasingly clear that in some instances limited human data will be required for a comprehensive assessment of the safety of new plant foods. This applies to both toxicological and nutritional assessment, and to the evaluation of allergenic potential. Such data may be obtained by means of volunteer studies and limited marketing, provided that comprehensive animal and laboratory tests have established that the food is safe enough for such purposes.

Some estimate of the safety of the food is necessary before human exposure; this should include, where appropriate, animal studies designed to reflect the nature and degree of the concern. The design of, and protocols for, these studies will be in part product-dependent.

4.4 Summary

Many of the potential hazards relating to the food safety of crop plants derived by means of the new biotechnology will only be expected to be present very infrequently. Even if these potential hazards are present, they will not pose any new risks as compared with those that might be expected from existing food sources and the genetic modification of crops via traditional plant breeding. Although the new technologies of genetic modification have prompted the present assessment of food safety, the concerns to which they give rise also apply to other forms of genetic modification in plants. In the light of the discussion in this section, a new paradigm for safety evaluation is recommended in which emphasis is placed on the characterization of the food and the use of the data to determine the need for appropriate toxicity tests.

5. Safety assessment of foods derived from animals generated by biotechnology

5.1 Introduction

Recent discoveries in genetics have made it possible to develop animal lines containing any desired sequence of exogenous DNA in the genome. Transgenesis is the process of introducing exogenous genes into the genome of cells or of newly fertilized embryos. The technology has been of scientific value in studies of gene expression, developmental biology and oncogenesis. Transgenesis may be able to provide a precise genetic route to developing farm animals that are disease-resistant, produce lean meat, or grow more efficiently.

Although not involving any genetic modification of animals, the production of purified protein pharmaceuticals, such as bovine and porcine somatotropins, which are identical or nearly identical to endogenous compounds, is mentioned here since it has given rise to certain concerns. Pharmaceutical products derived from biotechnological processes are evaluated in the same way as those made by other processes. While the effects of administered supplementary sources of endogenous hormones on the productivity and reproductive capacity of animals are still being considered, food safety problems have all generally been resolved (*15*). Furthermore, the use of endogenous substances exogenously administered has been thoroughly tested in the case of steroid hormones used to stimulate growth in cattle; these have not given rise to any major concern as to food safety in the international scientific community (*16*). While it may not be possible to regard all protein hormones that may be produced by biotechnology as belonging to the same class as the somatotropins, cost-efficient studies can be carried out relatively easily to check that they are safe.

5.2 Issues to be considered in safety assessment

Transgenic animals with gene sequences specifically coding for a desired gene product or characteristic can be reviewed from the standpoint of: (*a*) the safety of the gene product itself; (*b*) the consumption of the genetic construct; and (c) unintended effects of the inserted gene sequence.

5.2.1 Gene products

Gene products in the transgenic animal can cause both direct and indirect effects. The former result from the constant stimulation of receptor sites and tissues by protein products such as growth stimulants. As long as the specific protein concerned is known, food safety can be assessed in the same way as for exogenously administered material (e.g., injection of growth hormone). Constant elaboration of growth factors by the new genome could lead to disruption of the endocrine system and concurrent compensatory increases in endogenous steroid hormones which, in normal individuals, are effectively metabolized by the liver but may cause difficulties in individuals with hepatic insufficiency. However, this could lead to a problem only in unusual circumstances because the levels would need to be extremely elevated yet not detrimental to the animal.

5.2.2 Genetic construct

Consumption of the genetic construct representing new genetic material is, of course, a possibility with transgenic food-producing animals. There is little concern about the safety of consuming the gene itself. We eat the entire contents of animals, plants and bacteria. We also consume the genes of incidental contaminants associated with food products. Whatever DNA is eaten is degraded in the intestinal tract, and the increase in the purine and pyrimidine content of tissue resulting from the extra gene will be negligible compared with the total content of these substances.

The DNA construct will be of concern only if it is infectious, i.e., if it can be propagated in the environment or transmitted by the food to susceptible cells in the gastrointestinal tract. This is unlikely, since infectivity usually requires a specific viral protein to be present on the surface of the virus as part of the viral capsid or envelope. Some nucleic acids can infect mammalian cells in tissue cultures, but no mammalian diseases are known to be caused by the oral ingestion of coat-free nucleic acids.

Fully functional recombinant retroviruses could possibly emerge in the food animal after insertion by these methods. This would be likely to be attended by a viraemia and result in an animal in poor health and hence of poor productivity. New strategies of construct design limit infectivity and make the appearance of infective organisms unlikely. Many potential safety concerns regarding the use of retroviruses have been resolved with recombinant DNA technology.

5.2.3 Unintended genetic effects

Insertional mutations of host loci are possible as a result of the integration of foreign DNA within a host coding sequence. These mutations are likely to be discovered because the progeny will be suboptimal for food production. In addition, they make generation of homozygous lines difficult or impossible, and homozygosity is of significant advantage in animal husbandry.

Translocations and other DNA alterations can arise as a result of the insertion of a gene or the use of retroviral inserts. Such rearrangements might be associated with malignancies and, while not dangerous, would render a food product unsuitable for consumption if the malignancy developed during the relatively short life of the food animal concerned. The insertion of a transgene into a recipient genome is important from the point of view of safety because other genes producing potential toxins may be activated if they are linked to higher-level promoters. This is not a serious concern in healthy mammals because production of toxins that might also be harmful to humans will generally be incompatible with normal growth and development. Observations, veterinary examination, and postmortem inspection procedures should be capable of determining whether animals are affected by such toxins and, if so, the cause can be investigated. Furthermore, the genetic events causing the modulation of gene expression as a result of transgene insertion are not different from those that occur naturally. Records of animal breeding go back over 2000 years, and there has never been any indication that toxic lines of animals have been produced.

There is less certainty as to the absence of toxin genes in lower vertebrates and invertebrates (fish, shellfish and molluscs). While rare, seafood toxins, such as tetrodotoxin in puffer fish, are known. However, the fact that neither individual animals nor lines of animals of the common food fish have been found to produce a toxin, speaks strongly against the possibility of the activation of a toxin gene by a genetic insert. A test for toxin production may, however, be advisable in the more exotic species.

There are two other concerns. The production of new, more efficient, food animals must be monitored for any changes in nutritional quality or composition of the resulting food. The use of transgenic animals produced by both conventional and recombinant

DNA technology can have significant effects on the composition and nutritional quality of the food. These effects must be reviewed and assessed.

The expression of new genes creates the possibility that new food intolerances may arise. Possible problems can be envisaged with cross-species transgenesis and the production of heteropolymeric proteins resulting from the association of polypeptides from the transgene and the endogenous gene.

5.3 Safety assessment paradigm

New strains of animals and their products have not hitherto been routinely evaluated for wholesomeness and safety prior to human consumption. However, the consumer can feel reasonably safe, at least in the case of mammals, if the animals appear to be healthy. This is not necessarily true of fish and invertebrates, some of which produce toxins to which they are immune but which are harmful to humans.

In assessing the safety for human consumption of new strains of animals, the recommended approach is based on a combination of molecular, chemical and biological considerations.

Unpredictable effects and genetic rearrangements following the integration of a transgene may bring about changes in the levels of some of the normal constituents in transgenic animals. For those that have physiological consequences, the health of the animal will be a good indicator of the nature of the effect. In all cases, the effects of the gene product will need to be assessed. A thorough analysis of the genetic modification will indicate whether unpredictable effects are likely. Their impact will need to be evaluated, and the health of the animal, if a mammal, will be a good indicator of safety. Except in unusual circumstances, such as the formation of novel components, it will not usually be necessary to carry out safety tests on whole foods derived from animal sources.

5.4 Summary

While significant changes can occur as a result of the genetic modification of animal genomes, it would appear, on the basis of the current review of known or suspected hazards, that transgenic animals should not cause any significant concern from the point of view of food safety. It should be emphasized that, at least in mammals, food derived from a normally healthy and productive animal should generally be considered safe.

The exact gene product that may be the result of transgenic modification should be fully characterized as either an existing substance or one that may be new to the particular animal species concerned. The safety of gene products in food can be assessed in the same manner as for other animal drugs and food additives (*3, 16–18*).

In the safety assessment of food derived from transgenic animals, an appropriate system should be established for the review and evaluation of the molecular and chemical data.

Biotechnology can potentially result in significant changes in the nutritional quality of food and, in some instances, is specifically intended to alter the composition of food. The nutritional characteristics of conventional food products should therefore be assessed, since these are important in providing a basis for determining the significance of any changes that may be caused by biotechnology.

Food intolerances and food allergies may be a potential hazard, particularly with interspecies transgenesis or the production of new hybrid proteins. While these possibilities need to be kept in mind, it is highly unlikely that they will be a problem in the overall population. Current methods of assessing allergenicity might be used, although their serious limitations from the point of view of predicting a problem in sensitive individuals must be recognized.

6. Recommended safety assessment strategies for foods and food additives produced by biotechnology

6.1 Introduction

Any strategy for the safety assessment of foods or food additives produced by biotechnology will require the establishment of an appropriate framework for its regulation and enforcement. Biotechnology forms a continuum ranging from what have been termed the traditional biotechnologies to the new ones, and the issues and concerns relating to food safety are similar irrespective of the technologies employed. Concerns relating to individual products developed as a result of the application of individual technologies may exist but are specific to the product, not the technology.

The Consultation strongly recommended that any safety assessment strategy should be based on considerations of the molecular, biological and chemical character of the material to be assessed, and that those considerations should determine the need for, and scope of animal-based toxicological studies. This approach leads to a strategy for the evaluation of a product based on a knowledge of the process by which it has been developed, and a detailed characterization of the product itself. The Consultation considered that classical toxicity tests may have limited application in the safety assessment of whole foods and that, even for materials traditionally evaluated by these procedures, there is a need to review them with a view to developing a more mechanistic approach to safety assessment.

6.2 General considerations

The genetic modification of organisms by means of current technologies represents the latest point reached in a continuum of development rather than a unique branch of science. Many aspects of an assessment strategy are therefore common to all products irrespective of whether the method used to effect genetic modification is traditional breeding and selection, chemical or physical mutagenesis, or recombinant DNA technology. The elements common to the assessment of safety include a knowledge of the biological and molecular components of the system, and of the potential consequences of the modification, and the comparison of the final product with one having an acceptable standard of safety, usually the traditional product. For products without a traditional equivalent, wholesomeness will have to be established by means of accepted methodologies. The wholesomeness of new foods and food additives is determined in the same way as that of traditional foods and food additives, i.e., by testing for the presence of toxic constituents and impurities, and nutritional quality.

Where information on the biological, molecular and chemical characteristics of the product is insufficient to enable its wholesomeness to be assessed, there will be a need for tests in animal species. The type, scope and extent of the toxicological evaluation necessary will be determined by the adequacy or otherwise of the information already available.

6.2.1 Biological characteristics

The identity of the host and donor organisms should be established both genotypically and phenotypically. The organisms should be characterized with respect to known toxin production, relationship to known toxin producers in the same genus, and pathogenic, infective or toxigenic potential. The proposed use as food or pharmaceutical should be identified.

6.2.2 Molecular characteristics

Information on the technique used to develop the genetically modified organism, and an estimate of the possible consequences of the modification from the point of view of the wholesomeness of the food are necessary, and may include detailed descriptions of all the components used in the modification technique as well as the characterization of the transferred genetic material.

6.2.3 Chemical characteristics

The product should be characterized by chemical analysis, the scope of which will depend on the nature of the product, e.g., whether it is a purified substance having a simple chemical structure or a complex mixture. The analysis should be conducted with a view to comparing the analytical profile with that of the traditional food. Identification of new components will require additional safety assessment procedures, up to and including the isolation and characterization of the new components.

6.3 Specific recommendations

6.3.1 Safety assessment of genetically modified microorganisms and foods produced by them

1. Because of the diversity of foods and food ingredients derived from microorganisms, a large number of factors must be considered in assessing any potential risks in the light of the intended use of the substance in food.
2. The safety assessment should be based on sound, scientific principles and data and should be flexible so as to be able to accommodate scientific advances.
3. The approach to safety assessment should rely to the extent possible on the use of molecular, microbial, genetic, and chemical data and information in the evaluation of potential risks and the choice of appropriate safety tests.
4. General requirements in the safety assessment of food and food ingredients derived from microorganisms include the following:

    (a) the production organism and any organisms that contribute genetic material to it should be identified taxonomically and genotypically;

    (b) all introduced genetic material should be well characterized and should not encode any harmful substances; the modified organism should be genetically stable;

    (c) vectors should be modified so as to minimize the likelihood of transfer to other microbes;

    (d) selectable marker genes that encode resistance to clinically useful antibiotics should not be used in microbes intended to be present as living organisms in food. Food ingredients obtained from microbes that encode such antibiotic-resistance marker genes should be demonstrated to be free of viable cells and genetic material that could encode resistance to antibiotics;

    (e) pathogenic organisms should not be introduced into food. The modified production organism used to produce food ingredients should not produce substances that are toxic at the levels found in the finished product;

    (f) the safety of the modified production organism should be assessed with respect to the safety of the product of the

introduced genes (including allergenic effects when appropriate), the ability to alter adversely the nutritional composition of the food, and any appropriate biological containment.
5. When molecular, microbial, genetic, and chemical data establish that the food or food ingredient is sufficiently similar to its conventional counterpart, only minimal toxicological testing will generally be required.
6. The safety of foods and food ingredients derived from microorganisms depends on all the stages involved — strain development, production, processing, and purification. Each case must be evaluated in order to identify critical points and establish appropriate controls that will ensure safety and quality. Any change in the process should be evaluated in the light of these considerations. The maintenance of good manufacturing practices must be a fundamental part of any process.

6.3.2 Safety assessment of genetically modified plants and foods derived from them

1. The complexity of whole foods and the wide range of modification possible in whole foods derived from plants require an integrated approach to safety assessment, taking into consideration the proposed use of the food, the potential exposure, and the specific issues associated with the significance of the food in the diet.
2. The safety assessment should be based on the scientific principles identified as relevant to safety.
3. The safety assessment should be based primarily on a consideration of molecular, biological and chemical data.
4. General requirements in the safety assessment of foods derived from plants include the following:

   (a) the modified food crop and any organisms that contribute genetic material to it should be identified taxonomically and genotypically;

   (b) all introduced genetic material should be well characterized and should not encode any harmful substances. The modified food crop (e.g., the inserted genetic material and the target region in which it is inserted) should be genetically stable;

   (c) vectors should be modified so as to minimize potential transfer to other organisms;

   (d) the modified plant should not produce substances that are toxic at the levels found in the finished food product;

Recommended safety assessment strategies

(e) the safety of the modified plant should be assessed with respect to possible deterioration in the nutritional value of the consumed product.

5. The need for toxicity testing will be determined in part by the nature of the modified plant food. Molecular, biological, and chemical analyses should always be conducted before the need for animal testing is assessed. When the assessment of genetic and compositional change does not provide a satisfactory basis for the safety evaluation, it may be necessary to test the whole food in appropriate animal tests. The nature and extent of such testing must then be carefully assessed in relation to the need to provide additional assurance of safety.
6. Where a new constituent that has no history of safe use appears in a food, appropriate toxicity tests will be necessary.
7. After satisfactory completion of the safety assessment, where appropriate, the use of planned introduction or postmarketing monitoring will be necessary in order to address concerns not covered by traditional toxicity tests.

6.3.3 Safety assessment of genetically modified animals and foods derived from them

1. Characterization and establishment of the stability of the introduced genetic material should form the basis of the safety assessment.
2. The safety assessment should be based on sound, scientific principles and data and should be flexible so as to be able to accommodate scientific advances.
3. The approach to safety assessment should rely to the extent possible on the use of molecular, microbial, genetic, and chemical data in the evaluation of potential risks and the choice of appropriate safety tests.
4. Mammals are important indicators of their own safety, since adverse consequences of introduced genetic material will generally be reflected in the growth, development and reproductive capacity of the animal. The principle that healthy mammals only should enter the food supply is of itself a method of ensuring the safety of foods derived from animals.
5. Primarily because some fish and invertebrates are known to produce toxins, the healthy animal principle does not provide the same degree of assurance that food derived from such animals is safe and should be used with caution in determining the need for additional safety assessment.

7. Conclusions and recommendations

In its consideration of the food safety implications of the application of biotechnology to food production and processing, the Consultation reviewed its past, present and possible future applications. It examined the scientific principles that would need to be taken into account in assessing the safety of foods (including food ingredients, additives and processing aids) from microbial, plant and animal sources. The conclusions and recommendations of the Consultation are given below.

7.1 Conclusions

1. Biotechnology has a long history of use in food production and processing. It represents a continuum embracing both traditional breeding techniques and the latest techniques based on molecular biology. The newer biotechnological techniques, in particular, open up very great possibilities of rapidly improving the quantity and quality of food available. The use of these techniques does not result in food which is inherently less safe than that produced by conventional ones.
2. A number of food additives are already derived from genetically modified microorganisms. Food products derived from genetically modified plants are under development and are likely to be marketed in the near future. Although genetically modified fish and invertebrates may be available relatively soon, genetically modified mammals are likely to take longer to develop.
3. Biotechnological techniques can be used to prepare new, safer and more effective veterinary drugs, biopesticides, rhizobia, and other products for use in agriculture. By making it possible to develop highly specific reagents, biotechnology has also led to improved methods of food analysis.
4. Whenever changes are made in the process by which a food is made or a new process is introduced, the implications for the safety of the product should be examined. The scope of the evaluation will depend on the nature of the perceived concerns.

5. The evaluation of a new food should cover both safety and nutritional value. Similar conventional food products should be used as a standard and account will need to be taken of any processing that the food will undergo, as well as the intended use of the food.
6. Comparative data on the closest conventional counterpart are critically important in the evaluation of a new food, including data on chemical composition and nutritional value. The Consultation believed that such data are not widely available at the present time.
7. A new, multidisciplinary approach to the safety evaluation of new foods is desirable, based on an understanding of the mechanisms underlying changes in composition. Detailed knowledge of the chemical composition of the food, together with information on the genetic make-up of the organisms involved, should form the basis of the evaluation and will indicate the necessity for toxicity testing in animals. The approach will be facilitated by the integration of molecular biology into the evaluation process.
8. The Consultation agreed a set of scientific principles to be applied to the evaluation of the safety of foods produced by biotechnology, although at present they would need to be applied on a case-by-case basis.
9. In due course it will be possible to develop a framework for the evaluation of all new foods, including those produced by biotechnology. This will need to be flexible, the data needed depending on the nature and use of the product. There is at present little experience from which to develop general criteria for such a framework and, until such time as these criteria can be developed, a case-by-case approach is required.
10. As far as the products of the newer biotechnologies are concerned, detailed knowledge of their molecular biological properties will facilitate the evaluation process. It is already possible to identify many of the categories of data that will be necessary. In due course it will be possible to identify the genetic elements that are likely to be acceptable for use in food-producing organisms.
11. To facilitate the safety evaluation of foods produced by means of biotechnology, action at the international level will be necessary to provide timely expert advice in this matter to Member States of FAO and WHO, the Codex Alimentarius Commission, the Joint FAO/WHO Expert Committee on Food Additives and the Joint FAO/WHO Meeting on Pesticide Residues.

12. The Consultation concluded that, because of the rapidity of technological advances in this area, further consultations on the safety implications of the application of biotechnology to food production and processing will be advisable in the near future.

7.2 Recommendations

1. Comprehensive, well enforced food regulations are important in protecting consumer health, and all national governments should ensure that such regulations keep pace with developing technology.
2. National regulatory agencies should adopt the strategies identified in this report for evaluating the safety of foods derived from biotechnology.
3. To facilitate the evaluation of foods produced by biotechnology, data bases should be established on:
   — the nutrient and toxicant content of foods;
   — the molecular analysis of organisms used in food production;
   — the molecular, nutritional and toxicant content of genetically modified organisms intended for use in food production.
4. Consumers should be provided with sound, scientifically based information on the application of biotechnology in food production and processing and on the safety issues.
5. FAO and WHO, in cooperation with other international organizations, should take the initiative in ensuring a harmonized approach on the part of national governments to the safety assessment of foods produced by biotechnology.
6. FAO and WHO should ensure that timely expert advice on the impact of biotechnology on the safety assessment of foods is provided to Member States, the Codex Alimentarius Commission, the Joint FAO/WHO Expert Committee on Food Additives and the Joint FAO/WHO Meeting on Pesticide Residues.
7. FAO and WHO should convene further consultations at an appropriate time to review the Consultation's advice in the light of scientific and technical progress.

References

1. MULLER, W. H. et al. *Journal of molecular biology*, **124**: 343–358 (1978).
2. COHEN, S. N. et al. Construction of biologically functional bacterial plasmids *in vitro*. *Proceedings of the National Academy of Sciences of the United States of America*, **70**: 3240–3244 (1973).
3. *Principles for the safety assessment of food additives and contaminants in food.* Geneva, World Health Organization, 1987 (Environmental Health Criteria No. 70).
4. *Recombinant DNA safety considerations: safety considerations for industrial, agricultural, and environmental applications of organisms derived by recombinant DNA techniques.* Paris, Organisation for Economic Co-operation and Development, 1986.
5. HARLANDER, S. *Genetics of lactic acid bacteria.* Paper presented to the American Society for Microbiology Conference on Biotechnology, Chicago, IL, 7–10 June 1990.
6. VENEMA, G. et al. Competence and plasmid stability in *Bacillus subtilis* and prospects for the production of enzymes important for dairying in *B. subtilis*. In: *Proceedings of the ILSI International Seminar on Biotechnology, Tokyo, Japan, 9–10 June 1988.* Tokyo, ILSI, 1988.
7. MCKAY, L. L. Update on dairy starter cultures: genetics and biotechnology of dairy streptococci. In: *Proceedings of the ILSI International Seminar on Biotechnology, Tokyo, Japan, 9–10 June 1988.* Tokyo, ILSI, 1988.
8. GORBACH, S. L. Recombinant DNA: an infectious disease perspective. *Journal of infectious diseases*, **137**: 615–623 (1978).
9. WHO Technical Report Series, No. 789, 1990 (*Evaluation of certain food additives and contaminants*: thirty-fifth report of the Joint FAO/WHO Expert Committee on Food Additives).
10. PARIZA, M. W. & FOSTER, E. M. Determining the safety of enzymes used in food processing. *Journal of food protection*, **46**: 453–468 (1983).
11. GARDNER, M. C. Gastrointestinal absorption of intact proteins. *Annual reviews of nutrition*, **8**: 329–350 (1988).
12. VIHINEN, M. & MANTSALA, P. Microbial amylolytic enzymes. *Critical reviews in biochemistry and molecular biology*, **24**: 329–418 (1989).
13. EMORI, M. et al. Molecular cloning, nucleotide sequencing, and expression of the *Bacillus subtilis* (natto) IAM1212 alpha-amylase gene, which encodes an alpha-amylase structurally similar to but enzymatically distinct from that of *B. subtilis* 2633. *Journal of bacteriology*, **172**: 4901–4908 (1990).
14. LINDEMANN, J. Biotechnologies and food: a summary of major issues regarding safety assurance. *Regulatory toxicology and pharmacology*, **12**: 96–104 (1990).
15. JUSKEVICH, J. C. & GUYER, C. G. Bovine growth hormone: human food safety evaluation. *Science*, **249**: 875–884 (1990).
16. WHO Technical Report Series, No. 763, 1988 (*Evaluation of certain veterinary drug residues in food*: thirty-second report of the Joint FAO/WHO Expert Committee on Food Additives).
17. WHO Technical Report Series, No. 788, 1989 (*Evaluation of certain*

veterinary drug residues in food: thirty-fourth report of the Joint FAO/WHO Expert Committee on Food Additives).
18. WHO Technical Report Series, No. 799, 1990 (*Evaluation of certain veterinary drug residues in food*: thirty-sixth report of the Joint FAO/WHO Expert Committee on Food Additives).

Annex I
List of participants

Members

Dr K. Aibara, Head, Department of Food and Environmental Safety, Food and Drug Safety Centre, Hatano-Shi, Kanagawa, Japan

Dr A. J. Conner, Crop Research Division, Department of Scientific and Industrial Research, Christchurch, New Zealand

Dr L. A. Hussein, Department of Nutrition, National Research Centre, Giza-Dokki, Egypt (*Vice-Chairman*)

Dr W. R. Jaffé, Inter-American Institute for Cooperation in Agriculture, San José, Costa Rica

Dr D. A. Jonas, Biotechnology and Novel Foods Branch, Food Science Division II, Ministry of Agriculture, Fisheries and Food, London, England (*Rapporteur*)

Dr S. A. Miller, Dean, Graduate School of Biomedical Sciences, University of Texas Health Science Center, San Antonio, TX, USA (*Chairman*)

Dr N. K. Notani, INSA Senior Scientist, Biomedical Group, Bhabha Atomic Research Centre, Bombay, India

Dr A. Somogyi, Director, Max von Pettenkofer Institute, Federal Institute of Health, Berlin, Germany

Dr M. N. Volgarev, Director, Research Institute of Nutrition, Academy of Medical Sciences, Moscow, USSR

Representatives of other organizations*

Commission of the European Communities

Mr M. A. Granero Rosell, Foodstuffs Division, Internal Market and Industrial Affairs, CEC, Brussels, Belgium

EEC Scientific Committee for Food

Dr C. A. van der Heijden, Director of Toxicology, National Institute of Public Health and Environmental Hygiene, Bilthoven, Netherlands

International Food Biotechnology Council

Dr I. Munro, Director, Canadian Center for Toxicology, Guelph, Ontario, Canada

---

*Invited but unable to attend: International Organization of Consumers Unions; United Nations Industrial Development Organization.

International Life Science Institute

Mr M. Fondu, Director, ILSI Europe, Brussels, Belgium

Dr P. Niederberger, Vevey, Switzerland

Mr T. Takano, Biotechnology Working Group, Tokyo, Japan

Joint IUFOST/IUNS* Committee on Food, Nutrition and Biotechnology

Dr J. F. Diehl, Director, Institute of Nutritional Physiology, Federal Research Centre for Nutrition, Karlsruhe, Germany

Organisation for Economic Co-operation and Development

Dr V. Morgenroth, OECD Environment Directorate, Paris, France

Secretariat

Dr R. Carnevale, Assistant Deputy Administrator for Science and Technology, Food Safety and Inspection Service, United States Department of Agriculture, Washington, DC, USA (*FAO Consultant*)

Dr J. W. Gordon, Brookdale Center for Molecular Biology, Mount Sinai Medical Center, New York, NY, USA (*FAO Consultant*)

Dr J. Herrman, International Programme on Chemical Safety, Division of Environmental Health, WHO, Geneva, Switzerland

Dr F. Käferstein, Chief, Food Safety, Division of Health Protection and Promotion, WHO, Geneva, Switzerland (*WHO Joint Secretary*)

Dr I. Knudsen, Head, Institute of Toxicology, National Food Agency of Denmark, Söborg, Denmark (*WHO Temporary Adviser*)

Dr D. C. Mahon, Food Directorate, Health and Welfare, Canada, Ottawa, Ontario, Canada (*WHO Temporary Adviser*)

Dr J. Maryanski, Biotechnology Coordinator, Center for Food Safety and Applied Nutrition, United States Food and Drug Administration, Washington, DC, USA (*WHO Temporary Adviser*)

Dr Y. Pervikov, Microbiology and Immunology Support Services, Division of Communicable Diseases, WHO, Geneva, Switzerland

Dr J. Weatherwax, Nutrition Officer, Food Policy and Nutrition Division, FAO, Rome, Italy (*FAO Joint Secretary*)

---

*International Union of Food Science and Technology/International Union of Nutritional Sciences.

Glossary

Antisense RNA: RNA transcribed from the noncoding strand of the DNA of a gene, which is expected to form a complex with the RNA transcribed from the sense strand and inhibit translation.

Bacteriophage: a virus that lives in, and kills, bacteria.

Biotechnology: the integration of natural sciences and engineering sciences in order to achieve the application of organisms, cells, parts thereof and molecular analogues for products and services. (European Federation of Biotechnology, as endorsed by the Joint IUFOST/IUNS Committee on Food, Nutrition and Biotechnology, 1989).

Cosmid: a plasmid that can be packaged into phage particles and that, when used as a vector, combines the characteristics of plasmids and phages.

Electroporation: a technique in which foreign DNA is introduced into a cell under the influence of an electric current to improve transfer through the cell wall.

Food: any substance, whether processed, semi-processed or raw, which is intended for human consumption, including drink, chewing gum and any substance which has been used in the manufacture, preparation or treatment of "food"; does not include cosmetics or tobacco or substances used only as drugs.[a]

Food additive: any substance not normally consumed as a food by itself and not normally used as a typical ingredient of food, whether or not it has nutritive value, the intentional addition of which to a food for a technological (including organoleptic) purpose in the manufacture, processing, preparation, treatment, packing, packaging, transport or holding of such food results, or may be expected to result (directly or indirectly), in it or its by-products becoming a component of or otherwise affecting the characteristics of such foods. The term does not include "contaminants" or substances added to food for maintaining or improving nutritional qualities.[a]

---

[a] Codex Alimentarius Commission. *Procedural manual*, 7th ed. Rome, FAO, 1989.

Gene: a specific segment of DNA which generally codes for a protein required for a particular function of the organism.

Genome: the total hereditary material of a cell.

Microprojectile: describes a technique for introducing foreign DNA into a host on the surface of a small "bullet", which is fired into the recipient cell through the cell wall.

Pesticide: any substance intended for preventing, destroying, attracting, repelling or controlling any pest including unwanted species of plants or animals during the production, storage, transport, distribution and processing of food, agricultural commodities, or animal feeds, or which may be administered to animals for the control of ectoparasites. The term includes substances intended for use as a plant-growth regulator, defoliant, desiccant, fruit-thinning agent, or sprouting inhibitor, and substances applied to crops either before or after harvest to protect the commodity from deterioration during storage and transport. The term normally excludes fertilizers, plant and animal nutrients, food additives and animal drugs.[a]

Phage: *see* bacteriophage

Plasmid: (in bacteria) a small circular form of DNA that carries certain genes and is capable of replicating independently in a host cell.[b]

Promoter: a DNA sequence that is located in front of a gene and controls gene expression.[b]

Protoplast: the cellular material that remains after the cell wall has been removed.[b]

Quantitative trait loci: the locations of genes that together govern a multigenic trait, such as yield or fruit mass.

Recombinant DNA: DNA formed by combining segments of DNA from different types of organism.

Regulon: a protein, such as a heat-shock protein, that exerts an influence over growth.

Restriction enzyme: an enzyme that cuts DNA in highly specific locations.

---

[a] Codex Alimentarius Commission. *Procedural manual*, 7th ed. Rome, FAO, 1989.
[b] Taken from *Guidelines for the use and safety of genetic engineering techniques or recombinant DNA technology*, Washington, DC, Inter-American Institute for Cooperation on Agriculture, 1988.

Restriction fragment length polymorphism: the variation that occurs in the pattern of fragments obtained by cleaving DNA with restriction enzymes, because of inherited amino acid sequence changes in the DNA.

Retrovirus: an animal virus with a glycoprotein envelope and an RNA genome that replicates through a DNA intermediate.[a]

Safety cassette: (in this report) a piece of r-DNA attached to the desired gene, which is designed to eliminate traits in the recipient organism that might facilitate genetic transfer to other organisms.

T-DNA: (in this report) the segment of the Ti plasmid of *A. tumefaciens* which is transferred to the plant genome following infection.

Ti plasmid: a plasmid containing the gene responsible for inducing tumour formation.

Transposon (mobile element): a segment of DNA that can be transferred from one cell to another and be inserted at several sites in the recipient cell's DNA, with associated rearrangement of the recipient's DNA.

Zygote: a cell formed by the union of two mature reproductive cells.[a]

---

[a] Taken from *Guidelines for the use and safety of genetic engineering techniques or recombinant DNA technology*, Washington, DC, Inter-American Institute for Cooperation on Agriculture, 1988.